Management Division

Held at the 2024 AIChE Annual Meeting

San Diego, California, USA
27-31 October 2024

ISBN: 979-8-3313-1676-1

Printed from e-media with permission by:

Curran Associates, Inc.
57 Morehouse Lane
Red Hook, NY 12571

Some format issues inherent in the e-media version may also appear in this print version.

Printed with permission by Curran Associates, Inc. (2025)

For permission requests, please contact AIChE
at the address below.

AIChE
120 Wall Street, FL 23
New York, NY 10005-4020

Phone: (800) 242-4363
Fax: (203) 775-5177

www.aiche.org

Additional copies of this publication are available from:

Curran Associates, Inc.
57 Morehouse Lane
Red Hook, NY 12571 USA
Phone: 845-758-0400
Fax: 845-758-2633
Email: curran@proceedings.com
Web: www.proceedings.com

TABLE OF CONTENTS

INNOVATION AND ENTREPRENEURSHIP IN CHEMICAL ENGINEERING

600a Key Opex Steps Necessary for Growth – Set Yourself up for Success 1
Irene Legiec

600b The Lubrizol Sustainability Scorecard: A Comprehensive Tool for Evaluating and Communicating Sustainability in the Chemical Industry 2
Nicolas Proust, Glenn Cormack, Dixit Parmar

600c Michigan Climate Venture: A Multidisciplinary, Student-Led University of Michigan Program at the Intersection of Climatetech and Venture Capital 3
Shawn Lu, Gautam Kaul

600d Our Journey in Commercialising Technology for Producing Porous-Silicon for Li-Ion Battery Anodes 4
Gwen Chimonides, Siddharth Patwardhan

600e Gambling on Innovation and Experimental Portfolios 5
Darrell Velegol

LEADERSHIP AND MANAGEMENT OF SUSTAINABLE SOLUTIONS IN INDUSTRY

208e Lyten's Advancements in Lithium-Sulfur Batteries for Electric Vehicles 6
Kumar Bugga

208f Ucor's Practices Accelerate Progress Towards a Sustainable and Resilient Future 7
Brad Russell, Ethan Garland

208a New Resource for Sustainable Process Development 9
John Dever

208b Product Vs Process Sustainability: A Lubrizol Case Study 10
Michael DeLetter, Nicolas Proust, Glenn Cormack

208c Mission Possible – from Strategy to First Projects, Case Studies 11
Gayle Gibson

208d Environmentally Sustainable Process for Synthesis of Para Amino Phenol and Its Techno-Economic Benefits over the Conventional Process 12
Soukarsa Das, Arunava Maity, Pratyush Kumar, Abhijit Sarkar, Atanu Mukherjee

LEADING AND MANAGING INDUSTRY-ACADEMIA COLLABORATIONS AND PARTNERSHIPS

136d Improving Outcomes with Industry Input: How Engagement with Stakeholders Informs Science and Engineering 22
Peter Valdez

136b Academic-Industry Sabbaticals: A Professor's Perspective on Advancing Chemical Engineering Education 23
Ryan Hartman, Daniel Hickman

136a Academic-Industry Sabbaticals: An Industrial Researcher's Perspective on Sparking Innovation While Influencing Academic Education and Research 24
Daniel Hickman, Ryan Hartman

136e From Fortune 500 to Startup: <em>Perspectives on successful collaborations with Academia</em> 25
Ryan Lively, JR Johnson

136f Building Strong Industry, Academic, & National Lab Collaborations to Enable Technology Deployment: An ExxonMobil Perspective 26
Jason McMullan

PERSPECTIVES ON LEADERSHIP GROWTH FROM PREVIOUS MANAGEMENT AWARD RECIPIENTS

331a A Presentation by Past Management Award Recipient, Frank van Lier 27
Frank van Lier

331b Thriving as a Leader in a Rapidly Changing Environment 28
Senyo Opong

331c Sleepless Nights-hardest choices as a leader in industry 29
Gayle Gibson

331d Direct Air Capture-Leadership and Culture Matter 30
Robert Zeller

NON-TRADITIONAL CAREER OPPORTUNITIES IN CHEMICAL ENGINEERING FOR MBA'S, LEGAL PRACTITIONERS, AND PH.D.'S

270a From Lab to Legislation: A ChemE's Transition from Research to Regulatory Policy and Industry Advocacy 31
Owen P. Jappen

270b Using My Chemical Engineering Degree as a Patent Attorney 32
Michael Stephenson

270c Breaking Symmetry: Resilient Careers for Unpredictable Futures 33
Austin S. Lin

270d How a Young Technical Geek Wandered into Management and Later into Management at a Professional Society and Ended up Loving it. 34
Joseph Cramer

LATEST OUTLOOK AND KEY TRENDS SHAPING THE GROWTH OF CHEMICAL INDUSTRY (INVITED TALKS)

480b An Integrated Approach to Lead the Materials Industry through Disruption 35
David Parrillo

480c Current Trends and Opportunities in Pharmaceutical RandD 36
Shailendra Bordawekar

480e ExxonMobil Low Carbon Solutions: Accelerating the World's Path to Net Zero 37
Prasanna V. Joshi

480d Digital Transformation in the Chemical Industry using Advanced Process Modeling 38
Alejandro Cano

480a Leading Sustainability in a Dynamic Environment 39
Scott Collick

Author Index

600a Key Opex Steps Necessary for Growth – Set Yourself up for Success

Irene Legiec

As a small or medium size firm, you have an entrepreneurial mindset and are experiencing or anticipating a step change in growth. Now is the time to establish operational excellence (OpEx) and continuous improvement programs, to help streamline and supercharge your operations. Leaders will want to establish strategies that revolve around fostering innovation, implementing operational excellence, and ensuring sustainable growth.

As companies identify and implement innovative technologies to streamline operations, enhance productivity, and drive growth, consideration must take place on standardizing across the platforms. This will enable collaboration between departments and organizations and drive efficiency. Equipment and process standardization can simplify work processes, reduce errors, and can also streamline some processes for business growth.

Development of high-performance work teams to manage and run your business will be needed, including anticipation for staffing for growth and engineering optimization. What are the proper staffing levels for growth, while maintaining a lean organization? How do you keep employees engaged and trained? What is the ideal technical career ladder for your business? These personnel elements are critical as you map out your business model for growth and ensure employees stay connected with industry trends and develop the expertise needed to drive innovation within the organization.

Selection of key performance indicators (KPIs) are critical to track performance, and employees must be fully trained to track and evaluate the data. Now is the time to establish KPIs, train personnel, and track and measure the impact of innovation initiatives on business performance. Regularly monitor progress, solicit feedback from stakeholders, and iterate on innovation strategies to ensure continuous improvement and sustained growth.

Many innovative technologies and entrepreneurs will not have the bandwidth to develop these systems on their own. This presentation will provide a high-level overview of the steps towards Operational Excellence and highlight what is critical to establish now as companies continue to grow. This sets up the company for success, cuts waste, boosts quality and performance along the way.

600b The Lubrizol Sustainability Scorecard: A Comprehensive Tool for Evaluating and Communicating Sustainability in the Chemical Industry

Nicolas Proust, Glenn Cormack, Dixit Parmar

As the chemical industry transitions towards sustainable manufacturing, the need for robust tools to evaluate and communicate the sustainability impact of products and processes has become critical. This presentation will introduce the Lubrizol Sustainability Scorecard, an innovative tool designed to address this need by providing a comprehensive framework for assessing project sustainability metrics.

The Lubrizol Sustainability Scorecard enables the qualitative analysis and quantitative measure of the environmental, economic, and social dimensions of our products and processes. By integrating key performance indicators (KPIs) and utilizing a multi-criteria analysis approach, the tool offers a holistic view of sustainability performance. It can therefore be used to facilitate internal decision-making, strategy development, and project prioritization by providing clear, quantifiable data that can be easily understood and acted upon.

Externally, the Sustainability Scorecard could be leveraged as a powerful communication instrument. It could allow Lubrizol to transparently share sustainability data with customers, partners, and stakeholders, thereby demonstrating our commitment to decarbonization and sustainable practices. Such transparency could not only build trust across the company's network, but also help articulate the added value of sustainable products, potentially unlocking new market opportunities and driving competitive advantage.

This presentation will detail the development process of the Sustainability Scorecard, including the selection of relevant KPIs, the methodological framework, and selected case studies showcasing its application. Attendees will gain insights into how such a tool could be integrated into their organizations to foster their culture of sustainability and achieve their strategic sustainability goals.

600c Michigan Climate Venture: A Multidisciplinary, Student-Led University of Michigan Program at the Intersection of Climatetech and Venture Capital

Shawn Lu, Gautam Kaul

Founded at the University of Michigan in 2020, Michigan Climate Venture (MCV) is the first student-led impact investing fund with primary focus on decarbonization. Tackling the challenge of CO_2 emissions is one of the grand engineering challenges of the 21st century. MCV invests primarily in pre-seed and seed-stage startups with massive decarbonization potential in the Great Lakes region across three distinct verticals: food/agriculture, mobility, and materials/circular economy. Specifically, we provide patient capital which fills the "investment gap" between conventional grant funding and venture capital funding seeking market rate returns.

MCV places strong emphasis on deep partnerships outside of financial support – we work directly with our portfolio companies and provide *pro bono* consulting services to our portfolio companies to help them tackle one of the grand engineering challenges of the century. As a multidisciplinary fund of law, MBA, and PhD students, our core competencies include business strategy, sustainability, and technical/regulatory analysis.

In this talk, I will discuss three key focus areas for MCV:

- **MCV's partnerships with portfolio companies:** example projects with portfolio companies to evaluate business strategy, sustainability goals, and regulatory landscape analysis.
- **How MCV can partner with you:** MCV is always looking for exciting early stage decarbonization startups from the AIChE and chemical engineering community - we would love to hear about and learn more about your company!
- **Where MCV is going:** As a student-led fund investing in an ever-evolving sector, MCV seeks to stay ahead of the trend. We would love to hear chemical engineers' thoughts on pathways towards decarbonizing chemicals and other hard-to-abate sectors!

600d Our Journey in Commercialising Technology for Producing Porous-Silicon for Li-Ion Battery Anodes

Gwen Chimonides, Siddharth Patwardhan

We have developed an integrated methodology for accelerating the discovery to market journey for nanomaterials. This presentation will provide a background of this new approach and then discuss a case study of our recent discovery of a scalable, sustainable and safer method for producing high-value and stable porous silicon for battery anodes. It overcomes many issues of existing methods such as high production costs, problematic scale-up, unsafe manufacturing process and low-performance products.

Currently, we are undertaking steps towards commercialising this technology and, as such, performing

(i) process development,

(ii) product testing, and

(iii) market discovery.

In this presentation, we will share our efforts in the commercialisation journey. Special emphasis will be on market discovery, starting from understanding the commercial opportunity, exploring commercialisation options and developing routes to market.

600e Gambling on Innovation and Experimental Portfolios

Darrell Velegol

How to place bets on projects or experiments in your innovation portfolio? Innovation or capital budgeting portfolios can have dozens, 100s, even 1000s of projects or experiments. Companies use various methods to allocate their resources, including Stage Gate. While these processes provide guidance for the decision, they do not predict what should be the precise allocation for optimal returns at low risk.

Is there a framework that is practical and quantitative, that improves communication between the technical and commercial functions, that increases IRR and peace of mind, and which allocates your entire innovation portfolio in a global optimum, rather than just project-by-project or experiment-by-experiment? In this presentation I'll show a method that enables you to allocate rigorously the resources among all projects in your portfolio, using parameters for which you might already have estimates. I'll show examples of how this technique can increase IRR by 15% or more, improve technical-commercial communication and focus, and give you greater peace of mind as you make critical innovation resourcing decisions for your company.

208e Lyten's Advancements in Lithium-Sulfur Batteries for Electric Vehicles

Kumar Bugga

Among the 'Beyond-Li-ion' technologies, lithium-sulfur batteries have distinct advantages, including high specific energy, low cost, improved safety, abundant raw materials that are not plagued by supply chain issues, and low carbon footprint. Yet, their implementation has been challenging despite impressive results in academic environment, deterred mainly by the 'polysulfide shuttle' – a phenomenon attributed to the dissolution of polysulfides in electrolyte and formation of a redox shuttle. To address this, Lyten has developed a unique family of 3D Graphene materials from the cracking of hydrocarbons to produce hierarchical porous structure,[1] which has been reported to help mitigating the problem of polysulfide shuttle.[2] With its well-engineered nano/micro/meso porous structure, Lyten 3D graphene can physically retrain sulfur and the intermediate polysulfides. Furthermore, its mechanically flexible and electrically conductive framework counter the problems of volume change and poor conductivity of sulfur and polysulfides respectively, making it an ideal host for high-energy and long-life sulfur cathode in Li-S cells. In parallel, Lyten has been developing new cell components and materials, e.g., novel protected Li composite anodes including 3D architectures, advanced stable electrolytes that can function at low quantities (electrolyte/sulfur ratio) with adequate sulfation to support high rates of discharge, and multi-functional separators that can block polysulfide crossover to the anode. Integration of these advanced components resulted in Li-S cells with a specific energy comparable to current Li-ion cells (250-275 Wh/kg) and even exceeding for short-life applications. The cycle life is progressively improving, with the recent cells exhibiting 300 cycles at C/3 and 100% DOD and over 2200 cycles in a LEO (Low Earth Orbit) satellite cycling @ 20% DOD. There is expected to be further growth in the overall performance of Li-S cells, enabled by the ongoing tuning of the 3D graphene and advances in the cell components and materials, projected to be 325 Wh/kg and > 300 cycles by the end of this year. In parallel, various safety tests, including internal short through nail penetration, external short, overcharge and over-discharge and crush were performed on Li-S pouch and cylindrical 18650 cells, which showed impressive abuse tolerance without thermal runaway. Both the performance and safety results on the prototype cells have been corroborated by external 3rd party testing. Using their semi-automatic assembly lines (2.4 MWh) commissioned recently, Lyten has been producing cylindrical 18650 (and soon 21700) and pouch cells of 2-10 Ah, and recently shipped samples to various auto OEMs and the DoD customers.

[1] Lyten patent on "Carbonaceous material for Li-S cells" *US patent 11309 545 B2*, April 19, 2022.

[2] X. Ji, K. Tae Lee and L. F. Nazar, "A highly ordered nanostructured carbon–sulfur cathode for lithium–sulfur batteries", *Nature Materials*, Vol. 8, 500–506 (2009)

208f Ucor's Practices Accelerate Progress Towards a Sustainable and Resilient Future

Brad Russell, Ethan Garland

United Cleanup Oak Ridge, LLC. (UCOR), is committed to incorporating sound environmental management, protection and sustainability practices in all its work processes and activities that are part of its U.S. Department of Energy (DOE) contract. UCOR is committed to accelerating sustainable practices into daily activities. Addressing sustainability, carbon neutrality, and a circular economy are important aspects of Executive Order (EO) 14057, as well as DOE Order 436.1A which must be met through UCOR's contract with the DOE.

The UCOR environmental policy provides a framework for setting environmental objectives and includes a commitment to continually improve the environmental performance of operations. UCOR's commitment to protect and sustain human, natural, and cultural resources is inherent in its mission to complete environmental cleanup safely with reduced risks to the public, workers and the environment.

UCOR's Pollution Prevention and Waste Minimization program is a comprehensive and continuous effort to promote sustainable environmental stewardship by reducing the quantity and toxicity of all waste types and environmental releases, promoting environmentally preferable purchasing and resource conservation for UCOR activities across the Oak Ridge Reservation. Affirmative procurement, and specifically the purchase of recycled products, leads to a reduction in the environmental impacts of mining, harvesting, and other extraction processes and helps divert large amounts of materials from valuable landfill space while promoting a circular economy.

The UCOR Zero Emissions Vehicle (ZEV) Fleet Program Management Team has worked with a variety of internal and external stakeholders and partners to develop electric vehicle charging infrastructure. Accelerated ZEV initiatives resulted in the installation of electric vehicle charging stations at UCOR's main campus and an off-site shipping & receiving/central records storage facility.

Continued ZEV charging station expansion is an essential element in meeting the EO 14057 mandates for all light-duty vehicle acquisitions to be ZEVs by 2027. To accelerate ZEV fleet growth, in fiscal year (FY) 2024, UCOR's Fleet Management Team ordered ZEVs for all replacement eligible, light-duty vehicles. This includes 100 percent of all additional light-duty vehicles added to UCOR's fleet and an order for a medium-duty electric vehicle to expand the ZEV cleanup mission support capabilities. With UCOR's forward-looking ZEV plan, the designs for the new ZEV charging station were equipped with infrastructure to expand the capacity for additional chargers as new ZEV arrivals begin to outpace current charging capabilities. Additionally, given the nature of UCOR's cleanup work at various sites which frequently change location, the strategy includes the acquisition of a mobile electric vehicle charging station to promote ZEV program implementation at remote field locations to meet operational needs.

Partnerships with external stakeholders such as suppliers, community members, and the DOE enable access to resources, expertise, and broader networks, which fosters innovation, contributes to a circular economy, and increases UCOR's impact within the Oak Ridge Reservation and surrounding communities. Concurrently, engaging employees as active participants in sustainability efforts not only enhances corporate social responsibility but also cultivates a sense of ownership and commitment across the workforce. This is achieved through various on-site, sustainability centered events, educational opportunities, and internal newsletter articles. Through effective and meaningful employee engagement and external partnerships, UCOR's collaboration and internal mobilization drives meaningful progress towards environmental and social sustainability objectives.

With support from DOE and utilizing a top-down leadership approach for sustainability initiatives, UCOR uses strategies and frameworks aimed at fostering active participation and commitment from stakeholders across sectors. UCOR emphasizes the importance of inclusive and collaborative approaches to drive meaningful change, highlighting key initiatives that illustrate successful implementation. Collaboration between various stakeholders to advance workforce development and understanding of sustainability initiatives is imperative for the overall success of implementation. UCOR's practices aim to inspire action and innovation in accelerating progress towards a more sustainable and resilient future.

208a New Resource for Sustainable Process Development

John Dever

In 2022, the NIST sponsored development projects to address readiness against future disruptions to the U.S. supply chain due to pandemics or conflicts. RAPID, a Manufacturing USA Institute aligned to the process industries, was awarded a grant to support the development of methodologies to accelerate process technology development. The project created a national Center of Excellence for Process Technology Development (CoE) at AVN Corporation's location in South Charleston, WV that combines facilities, equipment and technical expertise to develop and demonstrate process technologies. The CoE is a new resource that supports the acceleration of Sustainable designs while managing costs and leveraging other resources. The CoE uses pilot scale modules that can be configured to develop and demonstrate sustainable processes and validate their metrics. The CoE successfully validated the modular pilot plant concept by demonstrating a decarbonized, continuous process to produce a pharmaceutical raw material.

This talk will focus on the opportunities to reduce development cost and time to market for Sustainable technologies through the use of flexible, modular pilot plants such as are available at the CoE.

208b Product Vs Process Sustainability: A Lubrizol Case Study

Michael DeLetter, Nicolas Proust, Glenn Cormack

Many Lubrizol products are built on feedstocks derived primarily from petrochemicals. To meet growing consumer demands in emerging natural-based markets, Lubrizol is exploring feedstocks derived from biological sources. Consumer products with environmentally friendly, natural credentials may pass the initial "green sniff test" but further evaluation using quantitative criteria defined in Lubrizol's Sustainability Scorecard can help deliver a process which is both natural and lower carbon intensity.

This presentation will review a case study in application of the Scorecard, using a novel development product. In this work, we will demonstrate how a holistic look at multiple angles of sustainability can lead to focused development in deficient areas to boot the overall positive impact of a product. We will also cover how the development team has resolved many technical challenges involved in scaling these types of products, including improving resource efficiency, reducing energy consumption, eliminating manual handling steps which reduce operator exposure, and evaluating of critical safety parameters. Additionally, the presentation will cover how use of natural based feedstocks have dramatic implications for the processing equipment required compared to their conventional, petrochemical contemporaries.

These efforts have ultimately demonstrated that naturally-derived labels do not equate to low-carbon labels on products, and only through thorough analysis and development activities can these two objectives be true simultaneously. Ultimately, this work will highlight the necessity of evaluating both the product (consumer facing) and the manufacturing process sustainability metrics within the Specialty Chemicals Industry.

208c Mission Possible – from Strategy to First Projects, Case Studies

Gayle Gibson

Sharing case studies of convening industry stakeholders to create strategy for hard to abate sectors and catalyze first projects to reach net zero by 2050. A couple sector examples will be shared and enabling technologies needed to accelerate the transition.

Environmentally Sustainable Process for Synthesis of Para Amino Phenol & Its Techno-Economic Benefits over the Conventional Process

Soukarsa Das, Arunava Maity, Pratyush Kumar, Atanu Mukherjee, Abhijit Sarkar*
Dastur Energy Inc.
111 Congress Avenue.,
Suite. 500
Austin, TX 78701
Phone: +15128230398
Email:Soukarsa.D@dasturenergy.com/ info@dasturenergy.com

Abstract

Para Amino Phenol (PAP) is considered to be one of the most important active pharmaceutical ingredients and a prime precursor of paracetamol. Approximately 60% of the global PAP production is utilized in the production of paracetamol, which is a widely prescribed analgesic and antipyretic. Another 25% of PAP produced, serves as rubber antioxidants, paints and pigments; 5% is utilized as a chemical intermediate; 5% in photography, while the remaining 5% finds application in different cosmetics and other miscellaneous uses. Currently, the global production of PAP stands at approximately 180-190 Kilo Tons Per Annum (KTPA). The global market for Para Amino Phenol (PAP) was valued at around 1.7 billion USD in 2023 and is projected to reach 2.5 billion USD by 2033, with a Compound Annual Growth Rate (CAGR) of 3.8 to 4% from 2024 to 2033. The key PAP demand centers are China and India, with 49 KTPA and 38.5 KTPA, respectively. China has an installed capacity of 100 KTPA PAP and is considered the market leader, accounting for over 60% of the global production. Traditionally, PAP has been produced through the reduction of p-nitrophenol using iron powder. This method involves the conversion of p-nitro chlorobenzene (PNCB) to p-nitrophenol, followed by its reduction to PAP. However, this process yields a low percentage of PAP, generates significant amounts of iron sludge and wastewater with chlorinated substances, and poses serious environmental concerns. Consequently, this method has been phased out in many developed countries. An alternative approach involves the hydrogenation reduction of p-nitrophenol using a Raney-Nickel catalyst. This method offers improved yield, fewer by-products, and reduced environmental impact compared to the traditional iron powder reduction. Nonetheless, it still presents the risk of environmental pollution due to the disposal of effluents contaminated with chlorine-based compounds like p-chloroaniline or p-chlorophenol. In light of environmental sustainability, a novel synthesis process using nitrobenzene as the raw material has emerged as a favorable alternative. This method employs catalytic hydrogenation of nitrobenzene in the presence of a Raney-Nickel catalyst to form phenylhydroxylamine (PHA) intermediate, which in acidic medium and in the presence of a Platinum-Carbon catalyst undergoes dehydrogenation to form p-amino phenol (PAP) in acidic medium. It offers a shorter, compact, and streamlined process route along with approximately 25% lower power consumption compared to the conventional PNCB route and ensures a satisfactory yield of PAP (above 83%) relative to the nitrobenzene feedstock at the dehydrogenation stage. Additionally, it produces value-added by-products such as aniline and a co-product, ammonium sulfate. This process ensures a high-quality product equivalent to pharmaceutical grade. Importantly, this approach eliminates the release of hazardous chlorinated compounds in effluents, thereby reducing environmental risks. A minor aqueous effluent is generated in this process, containing a low concentration (approximately 5%) of ammonium sulfate, which can be effectively treated before disposal.The production of PAP via catalytic reduction of nitrobenzene offers cost advantages, with approximately 40 to 45% lower conversion costs and around 30% reduced capital investment compared to the conventional PNCB route. This environmentally sustainable method, when implemented on a commercial scale with a capacity of 50 to 60 KTPA, yields over 85 to 90% at the determining dehydrogenation step, ensuring higher Earnings Before Interest, Taxes, Depreciation, and Amortization (EBITDA) and a higher Internal Rate of Return (IRR) compared to the conventional route over a plant lifecycle of 15 to 20 years. Considering the advantages from techno-economic and environmental aspects, this emerging route can be regarded as the future direction of PAP production. Our study provides a comprehensive comparative analysis, highlighting the advantages of this emerging approach over traditional methods from both techno-commercial and environmental sustainability perspectives.

Key Words: Para Amino Phenol (PAP), Active Pharmaceutical Ingredients (API), Nitrobenzene

Introduction

Para-aminophenol (PAP) serves as a significant industrial precursor in the production of pharmaceuticals, pesticides, dyes, and photographic materials. While the use of PAP in photography has nearly vanished with the rise of digital technology, its production is primarily fuelled by the synthesis of paracetamol (acetaminophen). This analgesic and antipyretic medication ranks among the most widely used globally, with annual production nearing 150,000 tons. From the reaction mechanism point of view, synthesis of paracetamol is the acylation of PAP as the final step, underscoring its essential role as an intermediate.

Historically, para-aminophenol (PAP) was synthesized by reducing p-nitrophenol using Bechamp conditions (Fe/HCl). However, this method produced substantial amounts of waste iron sludge and chlorinated byproducts, leading to significant environmental concerns. Subsequently, the production process for PAP shifted globally to involve the hydrolysis of para-chloro-nitrobenzene (PNCB), followed by catalytic hydrogenation of p-nitrophenol using Raney nickel as a catalyst. While this approach provides a relatively cleaner alternative, it still encounters issues related to the low selectivity of phenol nitration. Additionally, the formation of chlorinated compounds, such as p-chloroaniline and p-chlorophenol, along with small quantities of hydrochloric acid, continues to pose environmental risks.

The catalytic hydrogenation of nitrobenzene to produce phenylhydroxylamine (PHA), followed by the Bamberger rearrangement in acidic aqueous conditions, effectively synthesizes para-aminophenol (PAP) with a strong preference for the para position with around 85 to 90% selectivity. This process minimizes the formation of ortho-aminophenol (OAP), which is advantageous for achieving higher purity in the desired product. However, a notable byproduct of the hydrogenolysis of PHA is aniline. Additionally, when PHA undergoes catalytic dehydrogenation in the presence of a platinum-carbon (Pt-C) catalyst within an acidic medium, it leads to the formation of ammonium sulfate upon neutralization with ammonia. Ammonium sulfate is a valuable co-product with significant applications in various industries, including fertilizers and textiles. This entire process is particularly beneficial as it comprises relatively straightforward steps and effectively avoids the production of chlorinated byproducts in the waste streams. Furthermore, it offers a cost-effective method for the efficient synthesis of PAP, enhancing its appeal in industrial applications.

The demand for para-aminophenol (PAP) is largely driven by the pharmaceutical industry, where paracetamol stands out as one of the most widely used analgesics globally. This creates a significant need for PAP, especially in high-consumption regions like North America, Europe, & Asia. Currently, global production of PAP is estimated to be between 180,000 and 190,000 tons per annum. The market for PAP was valued at approximately $1.7 billion in 2023 and is projected to grow to $2.5 billion by the end of 2033, reflecting a compound annual growth rate (CAGR) of 3.8% to 4% during the period from 2024 to 2033. Key demand centers include China & India, with respective needs of 49 ktpa and 38.5 ktpa. China leads the production landscape with an installed capacity of 100,000 tons, supplying a substantial portion of global demand. In the United States, the demand for PAP is around 15,000 to 20,000 tons in 2023.

Supply factors are influenced by manufacturing capabilities, advancements in technology, and regulatory considerations that impact production processes. As the industry moves towards more sustainable synthesis methods to reduce environmental impacts, it faces challenges such as fluctuating raw material costs and competition from alternative pain relief medications, which can affect supply stability. Consequently, the interplay between supply and demand for PAP continues to adapt as industries prioritize sustainable and efficient production practices in light of environmental concerns and shifting market dynamics.

This paper presents a comprehensive overview of the cutting-edge technologies for producing para-aminophenol (PAP) from nitrobenzene, highlighting its techno-commercial advantages and contributions to environmental sustainability. By comparing these innovative methods with traditional production processes utilizing para-chloro-nitrobenzene (PNCB), the study underscores the significant benefits of modern techniques, not only in terms of efficiency and cost-effectiveness but also regarding their reduced environmental impact.

Key Objectives of the Study

The key focus of this paper is the environmental sustainability, better catalytic control, and commercial advantages of producing para-aminophenol (PAP) via the nitrobenzene route, compared to the conventional para-nitrochlorobenzene (PNCB) method. The novel process described employs catalytic hydrogenation of nitrobenzene followed by a dehydrogenation step to produce PAP, offering several advantages over traditional routes:

1. **Environmental Sustainability:** The nitrobenzene-based method eliminates the production of hazardous chlorinated by-products (such as p-chloroaniline and p-chlorophenol) found in the conventional PNCB route. This reduces environmental pollution

and simplifies waste treatment, contributing to a more sustainable production process.

2. **Improved Catalytic Control:** The use of a Pt-C catalyst in the dehydrogenation step ensures high selectivity for PAP, with minimal formation of side-products like aniline. This precise catalytic control leads to a higher yield of PAP (85-90%) over other isomers at the dehydrogenation stage and a more efficient process.

3. **Commercial Advantages:** The novel process offers approximately 30% lower capital investment due to a shorter process route and reduced equipment needs. It also results in a 40-45% reduction in production costs compared to the PNCB route. Additionally, the process generates valuable by-products, such as aniline and ammonium sulfate, which can be used in other industries (e.g., fertilizers and chemicals), further enhancing its commercial viability.

This emerging method provides a more environmentally friendly, cost-effective, and scalable alternative to traditional PAP production, making it a promising route for future industrial adoption. The details are depicted below.

Conventional Route for PAP Production from PNCB

The traditional hydrogenation reduction method for producing para-aminophenol (PAP) from p-nitrophenol also utilizes para-nitrochlorobenzene (PNCB) as the starting material and follows a similar hydrolysis and acidification process as the iron powder reduction method. However, this approach employs direct hydrogenation with Raney nickel as the catalyst, allowing for the reduction to occur in an aqueous solution. This method yields a higher output of PAP compared to the iron catalyst, generates fewer by-products, and minimizes waste liquid and residue discharge. Nonetheless, it still results in the release of co-products such as p-chloroaniline and p-chlorophenol, which pose environmental risks.

In this process, the hydrolysis reaction of p-nitrochlorobenzene with sodium hydroxide produces sodium p-nitrophenol. Subsequently, the acidification of sodium p-nitrophenol with hydrochloric acid yields p-nitrophenol. The final step involves the catalytic hydrogenation of p-nitrophenol in the liquid phase to produce crude PAP. The resulting product is then filtered, crystallized, centrifuged, and dried to obtain the final PAP product. An overview of the specific process flow is provided as follows.

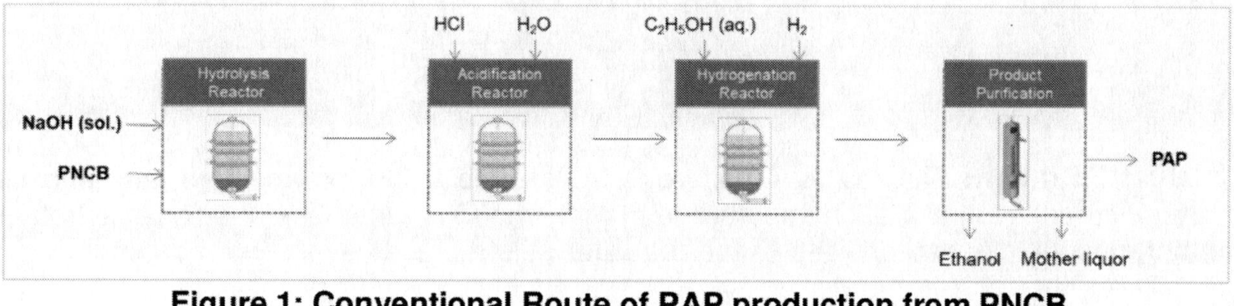

Figure 1: Conventional Route of PAP production from PNCB

A brief outline of the production process is as follows:

Hydrolysis Reaction: In this stage, an alkali solution and p-nitrochlorobenzene (PNCB) are transferred into intermediate tanks and subsequently pumped into a hydrolysis kettle. Nitrogen gas is injected to maintain pressure at 0.15-0.20 MPa. The steam jacket is activated to raise the temperature to 145 °C, after which it is turned off to allow the reaction heat to increase the temperature to approximately 165 °C. The reaction is maintained at this temperature for 2-6 hours, with a maximum pressure of 1.5 MPa. Using a 10% excess of sodium hydroxide, the reaction achieves a 99.5% conversion rate based on PNCB.

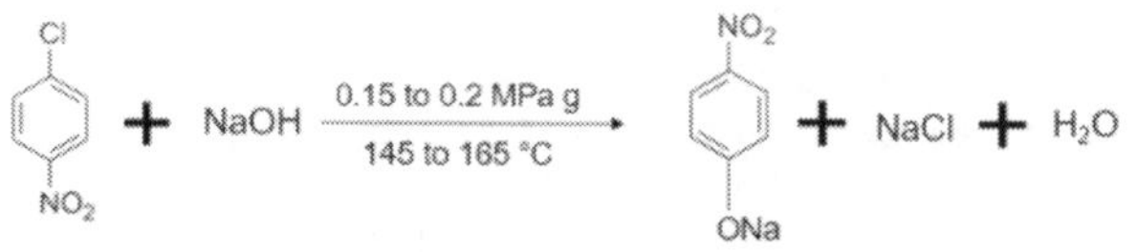

Acidification Reaction: Upon completion of hydrolysis, nitrogen is introduced to seal the sodium p-nitrophenol solution before transferring it to the acidification kettle. Hydrochloric acid is added slowly while the steam jacket maintains the temperature at 85 °C for 4-5 hours until the pH reaches 2-3. Following acidification, the mixture is cooled to room temperature for crystallization, lasting about 10 hours. With a 2% excess of hydrochloric acid, the reaction rate reaches 99.5% based on sodium p-nitrophenol. The resulting crystalline product is pressed into a centrifuge, while the mother liquor, primarily saline wastewater, is treated for recycling sodium p-nitrophenol.

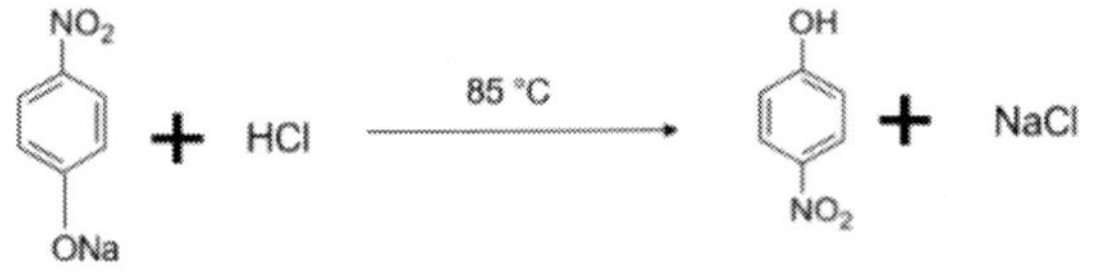

Hydrogenation Reduction: Fresh ethanol is pumped into a batching kettle and combined with water to create a 59-61% ethanol aqueous solution. Crude p-nitrophenol is added and stirred for about 0.5 hours until fully dissolved. A catalyst is prepared in a separate kettle with a 50% solid content.

Catalytic Hydrogenation: Before the reaction, nitrogen is introduced to create anaerobic conditions, followed by hydrogen gas. The catalyst solution and p-nitrophenol ethanol solution are added to the hydrogenation kettle, where stirring and heating (80-90 °C, ≤2.0 MPa) occurs for 1 hour. The reaction achieves a 99.5% conversion rate based on p-nitrophenol.

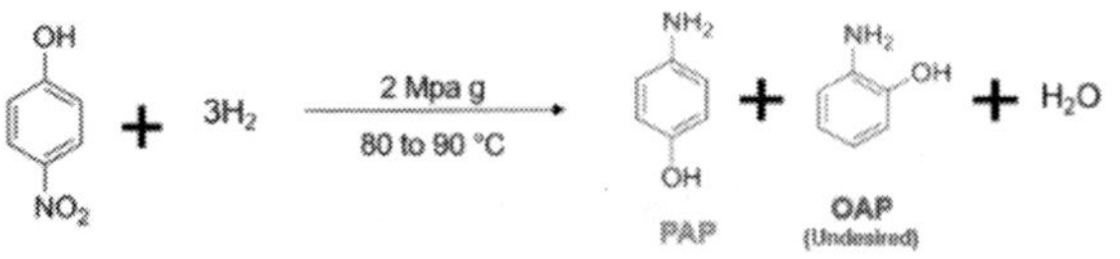

Catalyst Filtration: The reaction mixture is transferred to an intermediate tank and filtered through ceramic ultrafiltration membranes, with an 85% clear liquid discharge ratio. The membranes are cleaned bi-weekly, and the catalyst is replaced every 2-3 months.

Crystallization: The filtered solution is cooled to 0 °C to precipitate PAP crystals.

Centrifugation: The crude liquid is centrifuged, recycling some mother liquor and recovering ethanol for reuse. Crude PAP is sent to the soaking kettle.

Antioxidant Treatment: Crude PAP is soaked in a 15-20% sodium metabisulfite solution for 1-2 hours, with the solution reused and replenished as needed.

Drying: Wet PAP is dried at 50 °C in an airflow dryer and automatically packaged for sale.

Effluents and Emissions in the Conventional Production Process of Para-Aminophenol (PAP) from PNCB

In the conventional process for producing para-aminophenol (PAP) from para-nitrochlorobenzene (PNCB), several effluents are generated that pose environmental concerns. Key effluents include unreacted para-nitrochlorobenzene and by-products such as para-chloroaniline, which constitutes approximately 4-6% of PAP production capacity. This by-product is particularly threatening to the environment due to its toxicity. Additionally, the hydrolysis of PNCB produces chloride salts, including para-chlorophenol and various organic chlorinated intermediates, amounting to about 5-7% of production capacity. The process also generates corrosive hydrochloric acid, accounting for 3-4% of the production capacity, further contributing to the environmental impact of the process.

Emissions from the conventional PAP production route primarily include flammable hydrogen gas released during reaction processes and volatile organic compounds (VOCs) associated with chlorinated compounds like p-chloroaniline and p-chlorophenol. These VOCs present significant environmental and health risks, necessitating careful management and monitoring. Furthermore, the presence of chlorinated by-products in the effluents requires special handling to mitigate potential hazards. Overall, the conventional process generates various effluents and emissions that pose challenges to environmental sustainability and require effective mitigation strategies.

The Novel Process for the Production of PAP from Nitrobenzene in an Environmentally Sustainable Manner

This study outlines a novel, efficient method for the direct production of para-aminophenol (PAP) from nitrobenzene and hydrogen, utilizing concentrated sulfuric acid. This process is preferred for its simplicity, as it is a single-step operation that avoids the production of hazardous by-products and harmful exhaust gases, making it an environmentally sustainable alternative to conventional PAP production methods.

Key Steps in the PAP Production through the Novel Process

The process for producing PAP in this project consists of following steps:

- Hydrogenation of Nitrobenzene
- Addition of Acid to Phenyl Hydroxylamine (PHA)
- Catalytic Dehydrogenation of PHA
- PAP Extraction and purification Unit

Each of these units is designed to facilitate the controlled conversion of nitrobenzene to PAP through a series of well-managed reaction steps as illustrated in Figure 2.

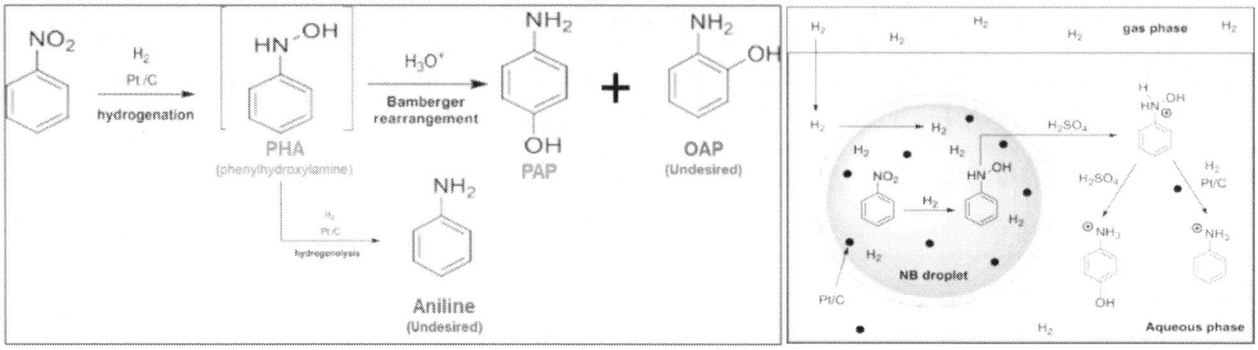

Figure 2: Reaction Mechanism of PAP production from Nitrobenzene

The novel method for PAP production begins with the hydrogenation of nitrobenzene, a key raw material. Nitrobenzene is continuously hydrogenated in the presence of hydrogen gas and concentrated sulfuric acid. This reaction produces phenylhydroxylamine (PHA) as an intermediate product. In the presence of sulfuric acid, the PHA undergoes a rearrangement to form para-aminophenol (PAP), with ammonium sulfate produced as a by-product. The process flow associated with this production route is as follows:

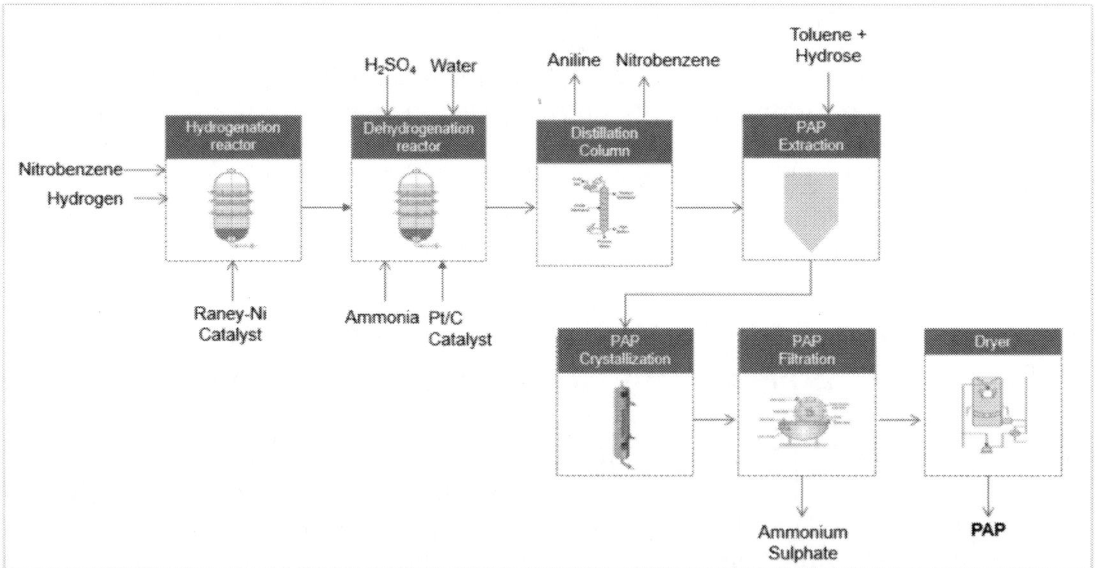

Figure 3: Novel Process Route of PAP production from Nitrobenzene

Hydrogenation of Nitrobenzene: The first stage of the process involves the hydrogenation of nitrobenzene to form phenylhydroxylamine (PHA). The feedstock, comprising nitrobenzene, toluene, and hydrose (which acts as a solvent), is continuously fed into the hydrogenation reactor. Raney-Nickel is used as the catalyst for the hydrogenation reaction. To maintain optimal reaction conditions, a constant flow of hydrogen gas is introduced into the reactor.

Nitrobenzene is added gradually to the reactor, allowing the reaction temperature to be controlled. The hydrogenation reaction produces phenylhydroxylamine (PHA) as an intermediate product. The reactor system consists of multistage reactors, each designed to ensure complete hydrogenation of the nitrobenzene feedstock. This stage is essential for producing high-quality PHA as the precursor for PAP.

Catalytic Dehydrogenation of PHA: After the hydrogenation stage, the PHA is subjected to catalytic dehydrogenation to produce PAP. This reaction is carried out in an acidic medium, with concentrated sulfuric acid acting as the catalyst, and is facilitated by a platinum-carbon (Pt-C) catalyst. The platinum content in the catalyst mixture is carefully controlled, typically ranging from 0.1% to 10% by weight. The catalyst mixture also contains a noble metal, such as zirconium sulfate, which enhances the efficiency of the reaction.

The dehydrogenation reaction involves two concurrent processes: the desired conversion of PHA to PAP, and an undesired isomerization reaction that produces aniline as a side-product. The platinum content in the catalyst plays a crucial role in ensuring high selectivity for PAP production. Excessive platinum can lead to a higher formation of aniline, which is an undesirable by-product. Therefore, the balance of platinum content is carefully managed to optimize the yield of PAP and minimize the formation of side-products.

Catalytic Control over the Success of the Dehydrogenation Process: The catalytic activity of Pt-C catalyst supported over Zirconium Sulfate for para-aminophenol (PAP) production during the dehydrogenation stage is characterized by a high turnover frequency of up to 500 h^{-1}. This process demonstrates impressive selectivity, achieving over 90% towards the para isomer in the dehydrogenation reaction. Remarkably, the catalyst maintains more than 80% of its activity after 5 to 10 reaction cycles, indicating its stability and reliability. Additionally, the catalyst features a specific surface area of 100-200 m^2/g, which enhances its efficiency. Overall, the dehydrogenation of PHA yields a high output of 85-90%, underscoring the effectiveness of this catalytic process.

Separation and Stripping: After dehydrogenation, the reaction mixture is sent to a stripping section to separate crude PHA liquor from unreacted nitrobenzene, toluene, and hydrose. The solvents are regenerated and recycled. Ammonia neutralizes sulfuric acid, forming ammonium sulfate, which is separated and treated.

Extraction and Crystallization: Solvent extraction removes impurities, and the mixture is cooled to crystallize PAP. The slurry is filtered, dried, and packaged as the final product. Residual liquids are distilled to recover solvents and aniline, leaving a slurry of ammonium sulfate, which is filtered, purified, and packaged.

Effluents and Emissions from the Novel Process Route for PAP Production

In the novel process for producing para-aminophenol (PAP) from nitrobenzene, several effluents and emissions are generated. The primary raw material, unreacted nitrobenzene, may be present in the waste streams, representing a potential environmental concern. Additionally, aniline, a valuable by-product of the process, is produced at a rate of approximately 7–8% of the total PAP production capacity. While this by-product could be captured for further use, its presence in the effluent streams necessitates careful management to mitigate any environmental impact.

A notable co-product of this process is ammonium sulfate, generated during the neutralization step. This co-product has a generation potential approximately similar to the PAP production capacity, making it a significant and valuable output that can be utilized in various applications, such as fertilizers. Regarding emissions, hydrogen gas, which is flammable, is released during the reaction processes. However, the process does not generate corrosive acids and chlorinated compounds thereby reducing the complexity of waste treatment and disposal. Overall, the innovative approach of this process aims to minimize harmful emissions and effluents while maximizing the production of valuable by-products.

Techno-Economic Advantages of the Novel Process over the Conventional Route

For the purpose of techno-economic (TE) comparison, this work has considered a PAP unit with 60,000 tons per annum production capacity. The TE comparison reveals around 30% benefit in capital cost of installation adopting the novel process compared to the conventional one. This is primarily due to the requirement of less equipment with a shorter process route and plot area requirement with novel process compared to the conventional one.

Moreover, utilization of the comparatively cheaper raw material and generation of by-products like aniline or co-products like ammonium sulfate makes the novel process more viable by achieving around 45% less cost of the production of PAP compared to the conventional processes. The comparative analysis of the Capex and the cost of production/ Ownership has been given in figure 4.

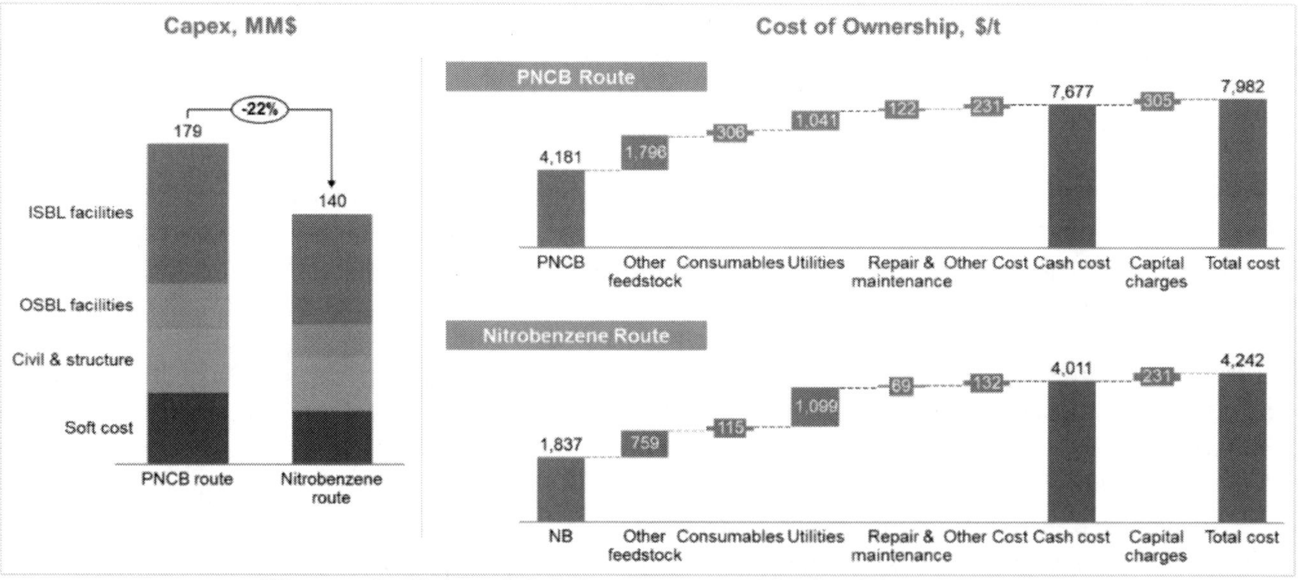

Figure 4: Financial Advantages with the Novel Route

Conclusion

This paper has reviewed the advantages of the novel process over the conventional process route from techno-economic and environmental sustainability aspects.

This novel process for PAP production from nitrobenzene represents a significant advancement in manufacturing technology. By employing a single-step hydrogenation and dehydrogenation process, the method eliminates the need for hazardous raw materials and effluent contamination with chlorinated by-products like p-chloro aniline or p-chloro phenol, which are commonly produced in conventional PAP production routes and thus eliminates the possibility of imposing strong and stringent statutory regulations for implementation of such production unit adopting the novel route.

Comparatively shorter process route gives a dividend in the plot plan requirement and capital cost of installation of the production units of similar capacities by reducing at least 30% compared to conventional process. The efficient use of catalysts and solvents, utilization of comparatively cheaper principal raw material combined with a focus on minimizing environmental impact, makes this process a sustainable and cost-effective solution for producing high-quality PAP at around 45% less production cost compared to the conventional one.

The integration of advanced process control and catalyst management ensures high product selectivity and maximizes the yield of PAP while minimizing the formation of undesirable side-products like aniline or o-amino phenol. Moreover the by-products like aniline and co-products like ammonium sulfate produced as a consequence of the novel process give substantial hedging to the overall product basket as these by-products and co-products find their own value-added utilization in some other associated sectors related to chemicals, textiles and fertilizer. This process not only offers an environmentally friendly and cost effective alternative but also provides a high-purity PAP product suitable for pharmaceutical and other industrial applications.

References

[1] Chinmay Ghoroi, et al. " Process Design and Economics of Production of p-Aminophenol"

[2] Manisha J Vaidya "Catalysis and Kinetics of Hydrogenation of Nitrobenzene to p-Aminophenol".

[3] Roxan, Joncour etc al. "Preparation of para -Aminophenol from Nitrobenzene through Bamberger Rearrangement Using a Mixture of Heterogeneous and Homogeneous Acid Catalysts".

[4] Market demand analysis by chemanalyst, marketsandmarkets and research by Dastur

136d Improving Outcomes with Industry Input: How Engagement with Stakeholders Informs Science and Engineering

Peter Valdez

Engaging with stakeholders throughout the research process is critical to success when developing new technologies and processes. Listening to stakeholder concerns can inform researchers about what topics and issues are most important and identify knowledge gaps or misconceptions. Outside expertise and knowledge can further enrich the research process resulting in a better final product than what may have been originally conceived. This presentation will highlight some best practices at Pacific Northwest National Laboratory for industry engagement. The presentation will also highlight formal collaboration mechanisms to consider when partnering with a national laboratory.

136b Academic-Industry Sabbaticals: A Professor's Perspective on Advancing Chemical Engineering Education

Ryan Hartman, Daniel Hickman

Chemical engineering academia has drifted in recent decades from its applied roots, perhaps unintentionally because of the separation between government as the primary funder and industry as the primary implementer of technology. Today's grand challenges in sustainability, circularity, and other spaces are large and complex, demanding the creativity and freedom of academic research partnered with the spirit of innovation in industry. Industrial sabbaticals are a unique learning opportunity for both professors and the host company to thrive. However, a successful sabbatical requires strategic planning to maximize the value of the experience. This talk will summarize a recent experience to offer professors helpful advice for designing and executing a successful sabbatical in a corporate research setting. Topics will include the choice of one semester or two, planning prior to the start, the avoidance of competing interests, the logistics of maintaining a university laboratory, educational opportunities within a company, and more. An industrial sabbatical can serve as a conduit between academia and industry to advance undergraduate and graduate education, both in the classroom and through the training of the workforce through academic research.

136a Academic-Industry Sabbaticals: An Industrial Researcher's Perspective on Sparking Innovation While Influencing Academic Education and Research

Daniel Hickman, Ryan Hartman

Some industrial researchers sense that academic research is drifting away from topics relevant to industry's greatest needs and opportunities. Industrial sabbaticals, in which a company hosts a professor onsite for several months, provide an opportunity to reverse this trend. These sabbaticals can create value for the sponsoring companies in several ways, such as infusing fresh ideas into their R&D communities, strengthening relationships with professors who can influence the academic curriculum and the training of future R&D employees, and accessing the expertise and creativity of professors to address current and future challenges. Based on recent experience hosting a 9-month sabbatical, this talk will describe ways to maximize the value of an industrial sabbatical. A successful sabbatical requires careful planning, such as identifying potential projects and collaborators, securing resources and internal funding, and ensuring clear communication and alignment of expectations. Finally, the sabbatical must also facilitate knowledge transfer at the end of the visit. By hosting industrial sabbaticals, companies can benefit from the unique perspective that professors can offer while influencing those professors' future research directions and teaching priorities.

136e From Fortune 500 to Startup: *Perspectives on successful collaborations with Academia*

Ryan Lively, JR Johnson

Collaborations with Academic partners – often referred to as "leveraging activities" around the water cooler – are an excellent solution to investigate challenging scientific and engineering problems. This approach allows the Company to expand into new areas of research when internal resources and/or time are limited. This talk will consider two case studies wherein two drastically different companies (by size and resources) collaborated with an academic partner, the Georgia Institute of Technology, creating a Venn diagram of overlapping success stories. The motivation for these engagements was driven by (1) expertise/talent in the space and (2) the capability to synthesize/fabricate/characterize materials at the University.

In case one, the Company – an oil and gas major – initiated a direct-funded project focused on novel membrane development for challenging hydrocarbon separations. This project quickly grew into a multi-year endeavor and included engagement across multiple disciplines. As the University produced new membrane materials, the Company would bring them in-house for rigorous performance evaluation using real feedstocks and conditions. Filling the pipeline from materials discovery to "real world" characterization proved quite successful.

As for case two, we consider the other end of the scale wherein the Company – a direct air capture startup – was smaller than the Academic research group at the time of engagement. Considering the Company resources (financial & physical) paired with the desire to expedite R&D, a different approach was required to initiate a meaningful collaboration. After the successful receipt of a government grant to fund the program, the Company and Academic team were able unlock the depth of capability in both materials/processes and characterization to accelerate scientific progress. The collaboration proved to be a springboard for the Company as it developed from ground up its in-house R&D capabilities and rapidly advanced the technology towards commercialization.

In this session, we will discuss these two industry-academia collaboration modalities, including lessons learned, key advantages, and frameworks for establishing successful collaborations. The speakers will split the allotted time to provide their individual perspectives having been the respective principle investigators for the projects in these two case studies. This will also provide a "mini-panel" format for Q&A after the presentation to serve various members of the audience.

136f Building Strong Industry, Academic, & National Lab Collaborations to Enable Technology Deployment: An ExxonMobil Perspective

Jason McMullan

Energy is essential for human development and improving living standards around the world. Society faces a challenge to provide reliable and affordable energy to meet the needs of growing population, while reducing environmental impacts, including the risks of climate change. Progress on society's energy and climate objectives requires practical approaches, new technology solutions, and collaborations and partnerships that enable human development and economic progress. ExxonMobil established its Low Carbon Solutions business in early 2021 focusing on carbon capture and storage (CCS), hydrogen and biofuels, leveraging our core capabilities in technology, scale, integration, and people. This presentation will provide an overview of energy fundamentals, ExxonMobil's strategy, and innovative approaches to technology development in the areas of CCS, Hydrogen and Biofuels.

Thousands of scientists and engineers, including more than 1,500 Ph.D.s, work at ExxonMobil. In R&D, they are exploring areas such as new catalytic and separation materials, novel low-energy process development and scale-up, advanced performance materials, and improved means of CO_2 storage. Our scientists have written more than 1,000 peer-reviewed publications and received more than 10,000 patents since 2010. In addition, we collaborate with more than 80 universities around the world, four energy centers, and several national laboratories. These collaborations have increased knowledge in key areas important to the energy transition: fugitive methane emissions detection and modeling; optimization techniques to understand CO_2 storage; electrification of processes; lower-emission fuels; and energy systems models.

331a A Presentation by Past Management Award Recipient, Frank van Lier

Frank van Lier

Dr. van Lier will speak on a topic of interest to a wide range of practicing chemical engineers. Dr. van Lier is a former recipient of the AIChE Management Award and a senior manager at Lubrizol with many years of both domestic and international experience.

331b Thriving as a Leader in a Rapidly Changing Environment

Senyo Opong

In today's fast-paced world, leaders are constantly challenged by evolving technologies, market dynamics, and shifting societal expectations. Thriving as a leader amid such rapid change requires a unique blend of adaptability, resilience, and visionary thinking. This talk will explore key strategies for navigating and succeeding in uncertain environments, focusing on the importance of continuous learning and innovation, and cultivating a flexible organizational culture. Attendees will gain insights into effective change management techniques, the role of emotional intelligence in leadership, and practical tools for maintaining team cohesion and morale during periods of uncertainty.

Dr. Opong is the recipient of the AIChE's 2021 Management Award and is Engineering Technology Leader for DuPont.

331c Sleepless Nights-hardest choices as a leader in industry

Gayle Gibson

Over a career in leadership, people can have sleepless nights from hard choices they face. What are those telling you about your leadership? Hear from an experienced corporate leader that will share learnings from some of the situations she experienced that included systemic dilemmas, ethical challenges, incongruence with company cultural norms and values and more. Takeaways include insight on ethical leadership and moral leadership and some frameworks that can be used when faced with difficult situations.

Gayle Gibson received the AIChE Management Award in 2018 for her work leading DuPont Engineering through the merger of DuPont & Dow and subsequent split into 3 companies.

331d Direct Air Capture-Leadership and Culture Matter

Robert Zeller

Oxy is making bold moves to commercialize Direct Air Capture (DAC) technology. DAC technology removes CO2 from the atmosphere and securely stores it deep underground or uses the CO2 as a feedstock. The UN IPCC has stated that DAC technology needs to be removing CO2 from the atmosphere at the multi-billion tonnes per year scale by 2050 in order to mitigate the worst effects of human-caused climate change. This presentation will highlight how leadership can set the culture to accelerate innovation.

Dr. Zeller received the AIChE Management Award in 2017 and in 2022 received the Winthrop-Sears Medal for exceptional innovation and entrepreneurship from The Chemist's Club.

270a From Lab to Legislation: A ChemE's Transition from Research to Regulatory Policy and Industry Advocacy

Owen P. Jappen

This paper explores the professional journey of a chemical engineer who transitioned from a technical career in research and development and process development/optimization to a chemical industry policy and scientific advocacy. Chemical Engineers, equipped with a strong technical foundation whether at the bachelor's, master's or doctoral levels, can combine their technical expertise and penchant for innovation with a motivation to influence broader industry practices and informed regulatory frameworks. The often interdisciplinary nature of chemical engineering, and unique viewpoint from industry experience, can accelerate and shape sustainable and scientifically informed chemicals management. This paper uses a case example of skills and experience to help other chemical engineers explore transitioning their own career, the impact of engineering expertise in the regulatory arena, and the landscape of chemicals management policy around the world.

270b Using My Chemical Engineering Degree as a Patent Attorney

Michael Stephenson

My name is Michael Stephenson, and I am an intellectual property attorney at the law firm Buchanan Ingersoll & Rooney PC, in Pittsburgh, Pennsylvania. Before going to law school, I attended the University of Pittsburgh and received a B.S. degree in chemical engineering in 2018. Until my senior year of undergrad, I fully intended on pursuing a traditional career as a practicing engineer. However, while working as a summer intern as a chemical engineering student, I started to have doubts regarding the career path that was best for me. I discussed my dilemma with a professor, and he mentioned that patent law may not only be a potential career path for engineers, but is often times a successful one. I took his advice and jumped straight into law school after receiving my engineering degree, and I graduated from the University of Pittsburgh School of Law in 2021.

After getting my law degree, I got a job at Buchanan Ingersoll & Rooney practicing intellectual property law. The vast majority of my practice involves patent prosecution, which generally relates to the process of obtaining patent rights. I get to utilize my chemical engineering knowledge on an everyday basis, as I have regular meetings with inventors to discuss their innovations and how they may be best protected.

During my presentation, I'll give you a more detailed description of how I got where I am now. I'll then provide more detail on what exactly I do at Buchanan and how I continue to use my chemical engineering degree in a non-traditional manner.

270c Breaking Symmetry: Resilient Careers for Unpredictable Futures

Austin S. Lin

It's not what you know, it's knowing what to do with what you know. In a business landscape where disruptive companies are themselves becoming disrupted or replaced, it's not sufficient to base career decisions only on title, salary and position. Instead of following the traditional career paths which mirror each other across industries, we explore nonlinear ways of strategizing your career, why applying the knowledge you have is more critical than only focusing on acquiring new knowledge alone. We'll focus on how to expand your career thinking, growing from short semesters to legacy-building decades and accelerating executive career experiences outside your present professional area to impact future opportunities. Your biggest success may be in a field that doesn't even exist yet. How will you prepare for industries that have yet to be created?

270d How a Young Technical Geek Wandered into Management and Later into Management at a Professional Society and Ended up Loving it.

Joseph Cramer

This presentation tells how I wandered into technical and project management as a young Ph.D. chemical engineer more or less by accident, certainly not by design. Later career developments forced me into a management role at AIChE. Although initially hesitant (actually scared to death and trying not to show it) I fell in love with the job and with the profession. My sound chemical engineering education proved to be extremely important throughout my long career in ways I could never have imagined.

480b An Integrated Approach to Lead the Materials Industry through Disruption

David Parrillo

Materials and chemicals impact daily life, as they enable modern healthcare, preservation of food, and delivering performance and efficiency of transportation and infrastructure. The need for sustainable solutions, including transforming waste into valuable resources and reducing the greenhouse gas emissions, is demanding new innovations and collaborations across industry partners. This talk will detail the technology advancements and cross-industry partnerships needed to transition an industry to one of low-to-zero carbon and circular solutions, driving innovations that will enable sustainable solutions for the world.

480c Current Trends and Opportunities in Pharmaceutical RandD

Shailendra Bordawekar

The pharmaceutical R&D pipeline has rapidly evolved over the last decade. Focus on development of new modalities for advanced therapies and challenges with complex drug candidates has driven adoption of novel concepts and technologies in pharmaceutical development. Innovation is key to reduce development costs and shorten time to market. Automation, predictive modeling via first principles and AI-ML are becoming increasingly important in the discovery and development of patient-centric medicines. This presentation will provide an overview of current opportunities and the application of these tools to drive pharmaceutical R&D.

480e ExxonMobil Low Carbon Solutions: Accelerating the World's Path to Net Zero

Prasanna V. Joshi

Energy is essential for human development and improving living standards around the world. Society faces a challenge to provide reliable and affordable energy to meet the needs of growing population, while reducing environmental impacts, including the risks of climate change. Progress on society's energy and climate objectives requires practical approaches, new technology solutions, and collaborations and partnerships that enable human development and economic progress. ExxonMobil established its Low Carbon Solutions business in early 2021 focusing on carbon capture and storage (CCS), hydrogen and biofuels, leveraging our core capabilities in technology, scale, integration, and people. This presentation will provide an overview of energy fundamentals, ExxonMobil's strategy, and innovative approaches to technology development in the areas of CCS, Hydrogen and Biofuels.

480d Digital Transformation in the Chemical Industry using Advanced Process Modeling

Alejandro Cano

Digitalization is revolutionizing the chemical industry by enhancing operational efficiency through advanced analytics, optimizing production processes, and reducing costs and environmental impacts. New business models are emerging, leveraging digital platforms for improved customer engagement and predictive maintenance. Additionally, digital tools are accelerating research and development, enabling faster creation of new chemicals and materials. High-fidelity digital twins combined with powerful computers and optimization algorithms now allow the systematic exploration of the process design space. Furthermore, there is a significant drive towards building autonomous process plants, which can operate with minimal human intervention, further enhancing efficiency and safety. Siemens is playing a crucial role in this transformation by helping their customers implement these autonomous solutions through co-creation and innovation partnerships, utilizing advanced process modeling tools to capture deep process knowledge, supplemented by real-time information available through communication with distributed control systems.

480a Leading Sustainability in a Dynamic Environment

Scott Collick

A chemical engineer's perspective on the sustainability challenges facing the planet today and the important role of product and process innovation. Sustainability has become a critical issue with significantly increased regulatory and stakeholder expectations. As we look to a more sustainable future, it's clear that technology will play a key role in driving global decarbonization efforts. This presentation will focus on addressing the world's most pressing sustainability challenges (including climate change, circularity, nature, water, and substances of concern) using science and innovation and how DuPont is working on lowering its footprint and expanding its handprint through its 2030 sustainability goals and its specific approach to each. Talk will detail how DuPont is managing sustainability across the enterprise including tools to evaluate the sustainability profile of products, how DuPont is managing increasing regulatory requirements, developing new products and solutions, and an effective management governance structure for sustainability to meet customer, investor, employee and other stakeholder needs.

AUTHOR INDEX

Bordawekar, Shailendra 36
Bugga, Kumar 6
Cano, Alejandro 38
Chimonides, Gwen 4
Collick, Scott 39
Cormack, Glenn 2, 10
Cramer, Joseph 34
Das, Soukarsa 12
DeLetter, Michael 10
Dever, John 9
Garland, Ethan 7
Gibson, Gayle 11, 29
Hartman, Ryan 23, 24
Hickman, Daniel 23, 24
Jappen, Owen P. 31
Johnson, JR 25
Joshi, Prasanna V. 37
Kaul, Gautam 3
Kumar, Pratyush 12
Legiec, Irene 1
Lier, Frank van 27
Lin, Austin S. 33
Lively, Ryan 25
Lu, Shawn 3
Maity, Arunava 12
McMullan, Jason 26
Mukherjee, Atanu 12
Opong, Senyo 28
Parmar, Dixit 2
Parrillo, David 35
Patwardhan, Siddharth 4
Proust, Nicolas 2, 10
Russell, Brad 7
Sarkar, Abhijit 12
Stephenson, Michael 32
Valdez, Peter 22
Velegol, Darrell 5
Zeller, Robert 30

AIChE
120 Wall Street, FL 23
New York, NY 10005-4020

ISBN 979-8-3313-1676-1

CURRAN ASSOCIATES INC.
proceedings.com